The
PC Builder's
Checklist

ANTHONY MOORE

ANTHONY MOORE

DEDICATION

To every toddler that puts their imagination and building blocks together to create something.
To those curious kids who take an interest in Science, Technology, Engineering, or Math.
To the nerd who steps to a different drummer in spite of the obstacles.

ANTHONY MOORE

CONTENTS

ANTHONY MOORE

INTRODUCTION

Congratulations! You're going to build a personal computer, or PC for short. Whether it's your first one or you are several "builds" in, it's always a great accomplishment. For anyone passionate about technology, there's nothing like finishing your "build", pushing the power button, and having it light up.

This book is not a comprehensive guide on PC building. It will not teach you how to build a system. It will not teach you about computer components. Nor will it explain diagnostics, troubleshooting, or even trying to guess what's wrong if your PC is not working. There are more than enough books on those subjects already.

This book does not go into component selection nor which company has the better products. Parts, products, and specs constantly change, with prices that can fluctuate between sites and within a matter of minutes. What you should buy is up to you and your budget to decide.

This book is merely a supplemental guide to assist you in getting your "build" from barebone parts to completion without forgetting any of the vital steps. It is a checklist that provides easy-to-follow steps with tips along the way, without all the fluff and filler. Because even the most seasoned builder can forget a step or two along the way when deep into a "build".

Anyone who attempts to use this checklist should already have some computer knowledge, including knowing a little about parts, some techno jargon, and at least be familiar with current PC standards. You don't have to be an expert, but this book probably won't help if you've never touched a motherboard before.

So, if you're ready. Turn the page and let's get started.

(Pssst… If you already know what kind of PC you want to build and have all your parts, you may want to skip the preface and go straight to page 1. Although you should check out the Appendixes at the back.)

On to the pre-check…

BUILD VS BUY

A computer is an investment of both time and money. When you need a system, you can go one of two ways to get it, build or buy. Before deciding, many people consider the advantages of both methods.

Reasons to Build Your Own PC

<u>Customization</u> - You can personalize your PC to your specifications and design choices.

When building a PC, you get to pick every component. You choose the case, CPU, GPU, RAM, storage, power supply, etc. You control how powerful it is and even the look—As long as you have the budget.

<u>Reusability</u> - If you already have a PC, you can usually reuse some components.

If you already have a "build" but need to upgrade, you can simply remove the old part then buy and install a new one. If your whole system is getting "up in age" then you may want to build a new PC and transplant some of the existing parts that are still useful.

<u>Economical</u> - Reusing or buying used parts can significantly reduce the cost of your PC.

Picking your own components means that you may have certain options in terms of pricing. If you're on a tight budget, then you might want to buying less expensive components. You may also be able to reuse or buy used parts— this can make the overall cost of your PC significantly lower than a pre-built system.

<u>Knowledge</u> – Building a PC is an exercise in learning.

If you've never built a PC, you will need to research the various parts and learn what goes where. You will be dealing with fragile components and must know how to put them together properly. Even a knowledgeable person must study, research, and learn what will and won't.

<u>Pride</u> - It just feels good to use something that you designed and built.

After just having completed putting together something that you've built with your own two hands, there is a justifiable sense of pride. Whether you've created your first bare bones "clunker", or if you've just "modded" the ultimate in PC art. There's no better feeling than to have made something yourself.

Reasons to Buy a Pre-Built

Time-saving - You don't need to invest several hours.

Building a PC requires time, including researching and ordering parts, shipping, unpacking, completing the build, cleanup, and testing. And that's assuming that you know how and everything you bought works correctly the first time. Should something go wrong, it takes time to diagnose and resolve the issues.

Stress – Building a PC can be stressful.

Despite the numerous resources available to newcomers and even seasoned builders, building a system can be stressful. There is always the chance that things can go wrong, especially if you don't have much or any experience. User error while building your PC could mean ruining expensive components.

User Error – Understanding the steps is crucial when it comes to PC building.

If you've never built a PC before, you must research the various parts and their proper locations. Even with guides and video tutorials at your disposal, you still may end up bending the pins on your CPU or snapping off the end of that irreplaceable connector.

Repairs - Most pre-built systems come with a seller warranty.

A pre-built system has a single warranty and a single company to stand behind it for a length of time. This ensures that you will have support and access to repair options if things go horribly wrong.

Simplicity - You can have the machine up and running in no time.

While building a PC can be an exciting endeavor, it's not for everyone. With a pre-built system, you don't need any technical knowledge or understanding of how components fit together. A pre-built system may also come preloaded with essential software. Some people just want a system that is simply plug-and-play.

Whether you build or buy, each options have pros and cons. It all depends on you and what you require of your PC. Building your own PC has always been considered the less expensive of the two. If you're an enthusiast who wants to customize your PC, then building your own system can be highly rewarding.

So, which one should you choose? You already know the answer. You wouldn't even have bought this book if you were planning on buying a pre-built PC.

PLANNING

The first part of accomplishing anything is planning. When building a PC, you should decide what will be the purpose of said PC. Will it be a business PC? Will it be a gaming PC? Will it be an editing rig? Will it be an Internet streaming PC? Or will it have some other far-out purpose? What do you really want or need it to do?

A business PC may need the fastest Central Processing Unit (CPU) available, with plenty of RAM, and can handle multi-tasking. A gaming PC may need the best Graphics Processing Unit (GPU) to handle intense parallel processing. An editing PC might need as much RAM and high-capacity storage as possible. An Internet streaming PC will need an exceptionally fast internet connection or even multiple network cards to handle a massive amount of data transfer.

It doesn't matter what your purpose is. The real issue is deciding on your PC's purpose and planning out the components needed to accomplish this before you ever touch a screwdriver. The internet is full of how-to guides, step-by-step videos, and forums to interact with other PC builders who can help with this phase of your "build".

BUDGET

Now that you've decided on a purpose, it's time to determine your **budget**! *GASP!!! PANT!!!*

YES! I said the B-word. How much do you intend to spend on your build? Or, how heavy is your wallet? Unless you have a rich father or uncle, your wallet is about to get a whole lot lighter.

PC building is not cheap. In the short term, buying a pre-built system can be cheaper than doing a "build", but manufacturers may use proprietary parts that cannot be upgraded down the line, eventually forcing you to buy a whole new system. Putting together your own system can save you money in the long run because you can upgrade your system piece-by-piece as new technology comes out, and you can afford better components.

Your next decision -- What kind of system do you want to build? Do you want a super high-end conquer all other systems type of PC? Or maybe a superior well-running performer? Or perhaps a mid-range working system? Or even a bottom-of-the-barrel wreck, with parts that may have been acquired through less-than-honorable means? You can't build a Ferrari on Ford money. You must figure out exactly how much computer you can afford.

The best thing to do is to figure out how much you have, then see how much you can squeeze out of your friends and family. Just kidding! Simply under budget and set aside the additional in case of price over-run. Finding a sale on outgoing/discontinued components or a discount for bundling (buying multiple parts all at once) can be a great way to stretch your budget. No matter how you do it, you don't want to hurt yourself by overspending. And I have a preventive cure for that.

Appendix A is a worksheet to assist you in calculating your budget. It can't give you accurate numbers, but based on general usage type, it can help you estimate how much you may want to spend on each component of your "build".

COMPONENTS

Sourcing parts is half the fun. If you aren't into technology, grab that geeky family member who is, treat them to lunch, and get their opinion. Workers at brick-and-mortar computer stores can be a wealth of information. I highly recommend checking out multiple manufacturer and reviewer websites.

Before purchasing anything, you should determine whether the components you intend to buy will be compatible, especially if you plan on installing used parts. Specifications change over time, and bleeding-edge components may not work with older parts. Used or older parts may not even have software or compatible drivers to enable them.

Compatibility is key. The motherboard you choose must be compatible with the CPU. You cannot buy an AMD CPU and then select an Intel motherboard and vice-versa. A cutting-edge CPU is not going to play well with a motherboard from three generations ago. Be sure to double-check for chipset and slot compatibility. You should also check for RAM compatibility with the CPU and motherboard.

Size is also a part of compatibility. GPUs, or video cards, have grown in size exponentially. Sometimes causing builders, upgraders, and modders (builders who modify the look of their builds) challenges when attempting to install them. CPU coolers have also gotten larger. Verify that your CPU cooler can fit on the motherboard with your RAM installed. Tall RAM sometimes has issues with larger coolers. This can become a major issue after you've purchased both and then attempt to install.

Since you roughly know how much each component will cost, selecting the best part at a particular price point should be easy. But if you find yourself in unfamiliar territory, there are multiple websites that can help you streamline your selection. Some even specialize in checking whether parts are compatible. It also doesn't hurt to check news articles and online reviews regarding different components. After all, you want to be happy with your parts once you've purchased them.

Once you've made your choices, figure out the final cost of everything plus tax and shipping if needed. Order and wait or go to the local brick-and-mortar and pick them up. Either way, you want to have most, if not all, of your parts and accessories in your possession before you begin your "build".

Appendix B is a parts planner where you can jot down multiple choices before making final selections.

Appendix C is a Power Supply planner that will help you determine how much wattage you may need before you make a final selection on which power supply to buy.

STORYTIME

My very first computer-related job was as a repair technician. I interviewed for the position with the manager and senior manager. I could tell the former liked me, but the latter had concerns. I hadn't had any formal training at that point, but I had "real world" experience. I answered all their questions but was not always able to use correct terminology. I explained that I had built several computers at home and networked a few of them for gaming purposes. I could disassemble, reassemble, and troubleshoot a PC. I believe that this is what convinced them to give me a chance.

In my first week as the newest tech, I was given my first repair call. The manager told me that it was a printer issue. The printer wasn't working, and a part needed to be replaced. He handed me the ticket and the replacement part. He said that it should fix the problem. Little did I know.

I showed up at the client's office. They showed me to the printer, and I did what I had been told to do. I replaced the part. I tested the printer, and it printed gibberish (technical term). I sat back and said to myself, "Time to think like a tech." Using the PCs to access the internet, I checked the support website for the printer manufacturer. I found out that they made two models of that printer, an older "A" model, and a more recent "B" model. The "B" model had newer software which was incompatible with the "A" model. I downloaded the "A" driver and installed in on their computer. I then tested the printer, and it worked just fine. The clients also tested it and were very happy. I informed them of how I fixed it and they signed off on the ticket. Repair complete.

Arriving back at the office, I was called into the manager's office. He and the senior manager were waiting. They asked if the printer was fixed. I said yes.They looked surprised. The manager asked if the part fixed it. I said, "No, there was nothing wrong with the printer." I got curious looks from both of them. I explained that the real issue was that the client was using an incompatible printer driver. I installed the correct driver and solved the problem. He asked where I got the correct driver. I replied, "The internet. I downloaded and installed it, and everything worked fine." They looked at each other, dumbfounded. The manager asked if I knew that five other techs had gone over there before me to fix the problem. I said, "No."

The point of the story is that even without any formal training, I came in, put my mind to use, and got the job done. Any new builder can do the same thing. So, never sell yourself short. It also illustrates that even several experienced persons can still be wrong. Sometimes all it takes is a fresh pair of eyes.

PARTS LIST

This parts list is a checklist of commonly used core PC parts necessary for building a complete system. Depending on the "build" you may have more components but I wouldn't advise having less.

- [ ] CPU – Brain of the system

- [ ] Motherboard – Heart of the system

- [ ] Memory – Stores the instructions for the CPU

- [ ] Cooler – Prevents the brain from overheating (Air, AIO, or Water)

- [ ] Fans – To cool everything else in the PC

- [ ] Storage – Where all programs and data are kept (Usually a HDD, SSD, or NVME)

- [ ] Power Supply – Powers this Frankenstein (Never cheap out on this part)

- [ ] GPU – Outputs video from PC to monitor (Some motherboards have basic video built-in)

- [ ] Case – Body of the PC which holds all the other parts

- [ ] Thermal Paste – Required to stick the cooler to the CPU for heat transfer

---

- [ ] NIC card – Usually built into the motherboard

- [ ] Wireless card – Some motherboards may have one built-in

- [ ] Sound card – Again, usually built into the motherboard

TOOLS

Having the right tools at your fingertips can make building a PC much easier. It can also keep a simple mistake from becoming an expensive nightmare. Most are necessary, but a few are just nice to have.

☐ Screwdrivers – Phillips #1 & #2, Flathead 3/8" or ¼". Screwdriver in kits, are magnetic, ratcheting, or electric are acceptable, but not advised because you don't want to overtighten or strip any screws

☐ Hex nut driver – For installing motherboard and serial port stand-offs

☐ Flashlights – Headlights and standing lights are acceptable

☐ Needle-nose pliers/Small vice grips – Good for removing screws, standoffs and zip ties

☐ Rubbing Alcohol – Used to clean the top of the CPU, especially if too much thermal paste is used

☐ Microfiber rag – To clean off thermal paste and absorb extra rubbing alcohol

☐ Screw Tray/Cups – Small cups for holding screws and standoffs while you are building

☐ Telescoping Magnet – At least one screw will fall into the PC where you can't get to it

☐ Cables – If you are using HDDs or 2.5" SSDs extra SATA cables and power cables are good

☐ Screws/Standoffs – A small collection of extra screws and standoffs can be a lifesaver

☐ Zip Ties/Velcro – These are useful for cable management and keeping wires from flopping around

☐ Antistatic mat – PCs should never be built while on carpet and/or wearing a sweater

☐ Antistatic wrist strap – A secondary precaution again static electricity which can kill components

ACCESSORIES

Accessories are useful items for the PC, but are not necessarily part of the build. You will need most of them to test and to actually use the system after completion.

☐ Monitor – Connects to the video card to produce the picture

☐ Mouse/Keyboard – Allows you to input data and move stuff around the screen

☐ Speakers – Connects to sound port to produce sound

☐ (2) 16GB USB Drives – One for installing BIOS updates and one for the operating system

☐ Ethernet cables – Connects PC to router, switch, or modem for internet access

☐ Power cables – Usually come with the case but have extra just in case

---

☐ Web camera – Necessary if you are going to do webcasts or participate in online meetings

☐ Microphone – Good for webcasts, optional if you have headphones or a mic in your webcam

☐ Headphones with Mic – Great for gaming

☐ Fan controller – Internal, used to control additional case fans

☐ RGB controller – Internal, use to control RGB case and/or fan lights

☐ Printer – Useful for printing out documentation and system specs

INSIDE - OUT

When building a PC, it's best to work on the inside first. Meaning it's best fit together the internal parts before messing with anything that touches the outside of the system. It's much easier to put components together outside of the case rather than trying to work within the confines of a small space.

IMPORTANT NOTE: A working PC with internet access will be needed before proceeding.

☐ Build a BIOS USB flash drive – Using a working PC, format a USB drive as FAT32. Download and copy the latest BIOS and drivers for your motherboard to the drive. A tool for flashing the drive may also be needed. Consult your motherboard manual and/or their support website for details. Write the BIOS version number on a peice of paper on the drive itself, and put it away. You will need this later.

☐ Build a USB Boot drive – Using a working PC, format a USB drive as FAT32. Download the operating system the build PC will use. This is usually in an ISO file image. Additional software may need to create the drive. This procedure is beyond the scope of this book and should be researched online before continuing.

☐ Verify both USB drives – Once you have both drives ready, verify them. Check that the latest BIOS is on the first drive and that the second one can boot the working PC. If successful, make sure that both drives are properly labeled. Put them away until the end.

☐ Start with a clean, dry surface – You will need a static-free area, such as a large table with enough space for you to be able to turn and move parts around. Rooms with rugs or carpeting are not advised.

☐ Use an antistatic mat and/or wrist strap - To prevent static electricity from building up and flash-frying your components. Ground your antistatic mat and/or wrist strap to dissipate any building electrical charge. If you don't have either, using the antistatic bag from the motherboard and building on the box as an ill-advised last resort.

☐ Unbox motherboard – Carefully remove the motherboard from all wrapping and set it down. In the box, there may be an I/O shield, a small aluminum bracket with a bunch of holes in it. Be sure to put this aside for the next part of the building process.

☐ Install CPU – **WARNING: Can damage the CPU and slot if done improperly.** Unbox the CPU, being very careful not to damage the connection points. Locate the locking arm next to the CPU socket and lift. Look for an arrow on the CPU. It points out the correct orientation. Align this arrow to the indicator on the motherboard. Carefully insert the CPU. It should drop into the socket without any force. Lower the locking arm to secure CPU.

☐ Install Cooler – **WARNING: Over-tightening the back bracket can damage the CPU and slot.**
 NOTE: If you're using an AIO (All-in-one, hybrid water cooler), you may want to delay completing this step until after installing the motherboard due to the need to install the AIO radiator first. Your cooler should allow enough space for the insertion of your RAM.

PRO TIP: Forgetting to apply thermal paste to the CPU can cause the CPU and motherboard to overheat and fry in a matter of seconds on power up.

Air / Tower Cooler

A) Unbox the cooler and verify if <u>thermal paste</u> has already been pre-applied to the copper plate. If not be sure to apply a thin EVEN coat to the top of the CPU.
(TIP: I use an old credit card to make sure I get an even coat.)

B) Many coolers come with a replacement back bracket for attaching the cooler to the motherboard. Check the CPU cooler installation instructions on whether to use the original motherboard bracket or to replace it with a new bracket.

C) Position the cooler over the bracket mounts and tighten each of the screws a few turns, one at a time, evenly until all are tight. Do not over tighten.

AIO / Water Cooling

A) Unbox the cooler (and fittings) and back bracket.

B) Replace back bracket with new bracket according to manual, if necessary.

C) Verify that <u>thermal paste</u> is already pre-applied to the water block. If not be sure to apply a thin EVEN coat to the top of the CPU.
(TIP: I use an old credit card to make sure I get an even coat.)

D) Position the water block on the bracket mounts. Be mindful of the orientation to in line of where the hoses, tubing, and/or radiator will be placed in the case.

E) Tighten the screws one at a time, a few turns at a time, evenly until all are tight. Do not over tighten, as this can damage the motherboard.

☐ Install the RAM – **WARNING: Too much force can damage the RAM Slot.**
 Carefully install RAM into the slots and push down until RAM clicks into place.

☐ Install Storage – If you are using an NVME drive, install the drive onto the motherboard and screw it down. If you are using an HDD or SSD, install into the cage or other space provided. Install the serial and power cables, if necessary.

PRO TIP: Install a GPU, then hook up a monitor, mouse, keyboard, and power supply outside of the case and test the system. Testing now saves disassembly time if something doesn't work.

OUTSIDE – IN

Now that the core components have been installed, it's time to prepare the rest of the system.

☐ Unbox the case – Remove the protection and put the accessories aside. Remove both side panels and set aside. Lay the case on it's side so that you can begin work.

☐ I/O Shield – Remember that <u>I/O Shield</u> that you set aside previously? This item is marked to tell you which ports are which on the motherboard. Install it now.

PRO TIP: If you don't install the I/O shield now, you will have to uninstall everything you've installed up to this point to have enough room to install. Then you'll have re-install everything again.

☐ Install motherboard standoffs – <u>Motherboard standoffs</u> that should have come with the case. Match the standoff holes in the case to the holes in the motherboard. Screw in the standoffs using your <u>Hex Nut Driver</u>. Double check that each standoff matches a hole.

☐ Install the motherboard – Carefully Install the motherboard into the case, matching the rear ports to the holes in the I/O shield. Align each motherboard mouthing hole to a standoff. Screw the motherboard down with PC screws using your Screwdriver #2.

☐ Install the radiator – If you are using an AIO cooler, install it now. You may have already connected the water block/pump to the CPU. If not, then you should also install it now.

☐ Install the case fans – **NOTE: Ensure your case fans are oriented correctly. Airflow only moves in one direction.** Air should come in the front (or bottom) of the case and exit the back (or top) to remove heat from the system. If you have a lot of case fans, you may need a fan controller because there may not be enough headers to power all the fans. If the fans contain also contain RGB, you may need an RGB controller.

☐ Install the power supply – **NOTE: Install the unit vent side down, if possible.** Set the case upright and install the power supply, securing it with the provided PC screws. If your power supply is modular, install only the cables needed to power your system components.

☐ Route your power cables – Modern cases have cutouts for routing cables behind the motherboard. Use these to route cables back into the PC as close as possible to where it belongs on the motherboard and plug it in. Use Velcro or zip ties to take up the cable slack and tuck it away neatly.

☐ Install storage – If you are using HDDs or SSDs, install them now. Most cases come with a cage or holder for drives, which may be removed if not being used. HDDs or SSDs will need serial ribbon or SATA cables and additional power cables. These must also be routed and plugged into the motherboard.

☐ Install front panel cable – Most cases have a front panel which has an internal cable, which must be routed and connected to the motherboard. The power button, reset, USBs, sound/headphone jacks, etc., are all controlled by this connection.

PRO TIP: This cable can have several very tiny wires that may have a specific orientation of positive and ground. Consult your case and motherboard manuals.

☐ Install add-on cards – Install any additional add-on cards, internal accessories, or peripherals, and secure them.

☐ Check the system – Go into the BIOS to ensure the motherboard works and recognizes all the installed parts. Check the speed and temperature of your CPU. You may want to enable your extended settings to speed up your RAM, check that all your storage space is recognized, and optimize the settings for your GPU.

PRO TIP: Testing now saves time and heartache later.

☐ Close the case – Set the case upright, grab the sidewalls, and reinstall them. If you have a clear side panel, it may be covered by a protective clear plastic sheet. Peel off that sheet.

☐ Update the BIOS – Remember that BIOS USB drive that you created at the beginning? Time to pull it out. Check the motherboard BIOS version against the one on the drive. If the one on the drive is newer, then time to update your BIOS. Most motherboards have a setting that allows updating the BIOS without an operating system.

WARNING: Losing power while updating/flashing a BIOS can brick your motherboard.

☐ Install the O/S – Time to install the all-important software that runs the entire computer, the operating system (or O/S for short). Grab the USB boot drive and boot the PC from it. Sometimes the BIOS is already set to boot from the USB, and sometimes you must press <F12> or <DEL> to force a menu, and select 'Boot from USB'. Once the drive comes up, it should prompt you to start installing the O/S.

☐   Install the drivers – Every component in your system requires and O/S compatible software driver. Each manufacturer has these drivers. Connect your system to the internet and download the latest drivers. Update all of them.

PRO TIP: Some motherboard manufacturers also provide <u>free</u> software for testing and tweaking your PC. I advise downloading and running it immediately.

CONGRATULATIONS!!!

You have completed your build. If this is your first build, you can consider this the foundation of your PC knowledge. To a new builder, if you didn't have any problems, it may not seem like you did all that much, but a veteran system builder knows just how important it was to complete these steps. A friend of mine once asked to watch me build a PC. So, we went to the store. I bought a bunch of components. We came back, and I sat down and put them together in about an hour. He was astonished. He said that it was so simple that even he could do it. And he very well could, but he couldn't see the downside.

Building a PC is great if everything works. But what if it doesn't? Even for a pro builder, trying to troubleshoot and diagnose the inner workings of a PC can be tricky. If a component isn't working, you can read the manual, try reseating it, and in a worst-case scenario, replace the defective part with a new one. Replacing new components under warranty can be time-consuming. Sourcing and buying replacement parts for used ones can be costly. Bad BIOS, drivers, and even software can cause any number of weird occurrences. One bad link can cause the entire system to fail. Therefore, testing is very important.

After your PC has been successfully tested, 'benchmarking' your system is a good idea. In other words, download software that will push your system to its limits to see how it does under pressure. You can also compare your systems to others online.

POST-BUILD

The final-final steps of PC building are ones that a lot of builders forget or don't even know about. Protection. Provision. And observation.

☐   Warranty – Each new component you buy will come with a warranty from the manufacturer.

Some used parts may come with a partial or remainder of the original warranty. I suggest that builders create a spreadsheet listing all your components, containing part numbers, serial numbers, places of purchase, dates of purchase, prices of parts, and lengths of warranty.
Appendix E provides an example of how to capture all the information. Registering your parts with the manufacturer may also be necessary. Not all companies require this additional step, but it's best to check. Some companies will deny a warranty claim because a part isn't registered.

☐   Anti-Virus – Protect your PC and all the information on it.

Even if you don't have any spare change left, your system will still need protection. You'll need to protect your system from worms, viruses, and trojans. Microsoft Windows comes with one type of Antivirus, but you may want something different. There are dozens of companies that sell protective software. Some of these companies may have "Freeware" versions of their paid product. In any case, no matter which product(s) you choose to use, you must keep it up to date to maintain that protection.

☐   Monitoring – Keep tabs on your system's performance.

Like a doctor's checkup, you should be keeping an eye on how your system is doing. Many manufacturers provide free software to monitor and maintain your system. A few of the major companies are famous for their software. A couple of Motherboard and GPU manufacturers even specialize in it. There are multiple third-party software companies that provide free monitoring tools. I suggest researching these tools, learning how to use them, and maintaining them.

APPENDIX A - Budget Worksheet

Filling out the worksheet below can be a good starting point. It's a chart of costs and percentages for different types of builds. Notice how a business PC and streaming PC can cost the same, but the emphasis on each budget is different because each PC has a different function and purpose.

Business PC - $800		Streaming PC - $800		Editing PC - $4000		Gaming PC - $1500	
CPU	20%	CPU	22%	CPU	13%	CPU	15%
MB	17%	MB	12%	MB	12%	MB	11%
GPU	15%	GPU	36%	GPU	37%	GPU	43%
RAM	4%	RAM	3%	RAM	18%	RAM	7%
Storage	24%	Storage	5%	Storage	4%	Storage	8%
Case	11%	Case	7%	Case	5%	Case	8%
PSU	6%	PSU	9%	PSU	5%	PSU	7%
Cooler	3%	Cooler	5%	Cooler	7%	Cooler	4%

NOTE: These numbers are not set in stone. If you want to get a cheaper GPU to upgrade the CPU, go right ahead. Just because you must go cheap now doesn't mean you can't upgrade later. If a new part comes out that's faster and better, then buy and swap. That's the beauty of building own your system.
Now get out your calculator because you have math homework.

INSTRUCTIONS: Pick a PC type and copy the percentages from the chart above. Write in your total budget, then use your calculator to figure approximately how much each part should cost.

TOTAL BUDGET: $_____

PC Part	Budget %	Calculated part cost (Total budget x Budget % = Approx. part cost)
CPU		$
MB		$
GPU		$
RAM		$
Storage		$
Case		$
PSU		$
Cooler		$

REMEMBER: These numbers do not account for the cost of accessories, extras, bells, whistles, or tax.

APPENDIX B – Parts Tree

Now that you know what kind of PC you want to build and approximately how much each part will cost, it's time to start making some real decisions. Compatibility is key. There are whole websites with tools dedicated to helping you choose compatible options that can fit within your budget.

INSTRUCTIONS: Circle your choices on the tree and write down one to three alternatives under it. Remember, your first option may be sold out or unavailable. Some of your choices may have multiple versions or models available. These choices are yours to make based on your needs versus your budget.

CPU:
AMD or INTEL

Motherboard:

Basic	Mid-range	Expert	Pro	Ultimate

CPU Cooler:

Air	AIO	Water
Fans vs Heatsink	120mm, 240mm, 360mm	Hard vs Soft lines

RAM:	Storage:	PSU:
Capacity vs Speed vs Overclock	HDD vs SSD vs NVME	Modular vs Semi vs Non

GPU:
AMD vs NVIDIA vs INTEL

Basic	Founders	Partner	Overclocked	Super/TI

Case:

HTPC/Small	Mini-Tower	Mid-Tower	Full-Tower	Super/Ultra

Case Fans:
Number and size

APPENDIX C – PSU Wattage Worksheet

Now that you know the parts you plan on using, one last thing is of vital importance, your power supply. Unfortunately, this is one of the most overlooked parts of building a PC. Some builders skimp on this part to save money for others. Is this a good idea? NEVER!!! If anything, you should be looking to move money from one of your other components and buy a good quality power supply. It provides and regulates the power to your very voltage-sensitive CPU, GPU, and motherboard.

Every component in your PC requires power, so your power supply must be beefy enough for use. Filling out this worksheet can provide you with a rough estimate of how much power your system may need. Buying a power supply with a higher wattage than you need is good, but you should also make sure it has a decent efficiency rating.
(This area is beyond the scope of this book, but further information is available online.)

INSTRUCTIONS: Check the power specification for each of your components. Write them in on the worksheet below, total them at the bottom, and round up to the nearest 10. This will give you a rough estimate of how powerful a power supply you will most likely need to buy.

CPU: (250w – 400w) _____

GPU: (80w – 350w) _____

Motherboard: (30w-80w) _____

CPU Cooler: (1w – 2w) _____

RAM: (3w – 5w) _____

Storage: (2w – 4w / per drive) _____

Webcam: (2w – 4w) _____

TOTAL WATTAGE _____

APPENDIX D – Quick Build Summary

Now that you've finished your build, here is a build summary that can be ran through quickly to ensure that all steps have been taken.

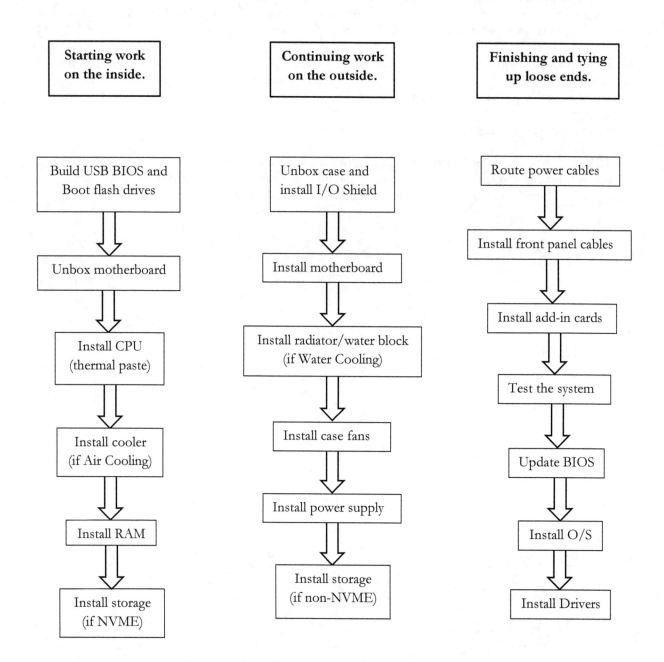

APPENDIX E – Warranty List

Even if your build runs successfully, you may need an RMA (Return Material Authorization) to send back a component later. In other words, your PC may be running fine and suddenly...BAM!!! Something breaks, and your PC doesn't work, and you may need to return a broken part. If the component is covered under the manufacturer's warranty, you can send it back for a repair or replacement.

The first step is documenting the information for your parts. The second step is registering those parts with the manufacturer. Not all companies require the second step, but it's best to check their website and find out their policies. Some companies will deny a warranty claim if a part isn't registered online in advance.

Warranty List Example

(Unless you can write really small, I'd advise creating your own.)

Manufacturer	Part/Model #	Serial #	Place of Purchase	Date of Purchase	$ of Part	Length of Warranty	Notes

ABOUT THE AUTHOR

I'm Anthony Moore, a native of Chicago, and a military veteran with almost four decades of experience with computers. I've done builds, hardware repair, software installation, custom scripting, database work, networking, administration, and more. If it has to do with computers, I've probably touched it. I'm a self-taught Jack-of-all-trades, expert of none.

I built my first computer at age 17, so that I could play "DOOM!". I spent days finding, obtaining, and putting together parts. Once the "build" was finished, I pressed the power button, the system lit up, and it didn't work. At that point, I didn't know that I needed to install an operating system. In fact, I didn't even have one. Fortunately, this was way back in the day when there was DOS and it fit on a floppy disk.

I went down to my local Sears (the Walmart of yester-year) which had working computers on display. I popped in a floppy disk, copied the entire DOS directory, took it home, copied it to my PC, rebooted, and waited. Nothing happened.

This is where I had to go pick up a book and learned about the hidden files which allow DOS to boot. After another trip to Sears and copying the hidden boot files, I returned home, installed the files, and WHALLA!!! A working PC. After installing and booting up the game, I discovered that still needed a sound card. (This was before motherboard manufacturers built sound chips into their products.) Nevertheless, I was eventually able to happily waste hours and hours playing the first level of "DOOM!".

Sounds bad, but that game provided a 17-year-old kid with the motivation and drive to learn something new and complete his first "build". I gained a new skill and a passion for technology that I might otherwise have missed growing up. A passion that I hope to pass onto others.

CONTRIBUTORS

This page is to acknowledge all 'royalty free' artwork and their creators used to build this checklist.

Cover – Information Wall Background created by SashaSan – Pixabay.com
   Clipboard created by bluebudgie – Pixabay.com,
Page 5 – Tile icon created by Clker-Free-Vecor-Images – Pixabay.com
Page 5 – USB vector icons created by OpenClipart-Vecors – Pixabay.com
Page 5 – Table icon created by Clker-Free-Vecor-Images – Pixabay.com
Page 5 – Power by Hat-Tech PK – thenounproject.com
Page 5 – Motherboard icon created by Chattapat – Flaticon.com
Page 6 – CPU icon created by Denis Shumaylov RU – thenounproject.com
Page 6 – CPU Cooler icon created by Chattapat th – thenounproject.com
Page 6 – Liquid Cooler 240 created by KEN111 th – thenounproject.com
Page 6 – RAM icon created by Mahmure Alp TR – thenounproject.com
Page 6 – SSD icon created by AB Design – flaticon.com
Page 6 – Hard Disk icon created by Memed_Nurrohmad – Pixabay.com
Page 7 – Box icon created by OpenClipart-Vecors – Pixabay.com
Page 7 – Hexagon icon created by OpenClipart-Vecors – Pixabay.com
Page 7 – Server icon created by OpenClipart-Vecors – Pixabay.com
Page 7 – Stop icon created by Clker-Free-Vecor-Images – Pixabay.com
Page 7 – Case Fan icon created by Smashicons GB – thenounproject.com
Page 7 – Power Supply icon created by ferdizzimo TH – thenounproject.com
Page 7 – Traffic Sign icon created by CopyrightFreePictures – Pixabay.com
Page 7 – SATA icon created by Arthur Shlain – thenounproject.com
Page 8 – Computer Wires icon by Pike Picture BY – thenounproject.com
Page 8 – LAN Card icon by Alina Oleynik - thenounproject.com
Page 8 – Checklist icon by Claire Dela Cruz – Pixabay.com
Page 8 – CPU Chip icon by Sinisa Maric – Pixabay.com
Page 8 – Windows icon by Raphael Silva – Pixabay.com
Page 8 – Penguin icon by Lars Meiertoberens DE - thenounproject.com
Page 9 – Quality icon by Wahyono Budiargo ID - thenounproject.com

---

www.ingramcontent.com/pod-product-compliance
Lightning Source LLC
La Vergne TN
LVHW060039070326
832903LV00072B/1384

* 9 7 9 8 9 8 8 3 5 9 9 0 6 *